This

BOOK

BELONGS TO

_____

_____

DEDICATION

This book is dedicated to all the amazing bird watchers around the world!

You are my inspiration in producing books and I'm excited to help in the planning and tracking your birding adventures around the world!

This Bird Watching Journal is a perfect way to capture all your birding thoughts, report sightings and keep track while you're out watching birds. Easily keep all your birding information in one place.

Each page is guided and has prompts that include:

1. Season - Which season am I looking for birds. Winter, summer, fall, spring.

2. Date/Time/Location- Record the date, time you arrived, where you are located.

3. Weather/Elements - Record what the weather is like and any important elements while out bird-watching.

4. Place/Location/Habitat - Space to write the place you decided to visit to look for birds and describe your location, favorite places for watching birds, and special notes on the habitat of your favorite birds.

5. Sights/Sounds/Activity - Note space to write down your surroundings, what you see, what you're hearing, any activity around you.

6. Bird species/Markings/Features- Keep Track your observations of bird size, tail, range, color pattern, and behaviors of the species of bird you are watching for.

7. Notes - Extra note space for favorite moments or highlights out in the field, things to watch for next time or just a summary of visit.

8. Whether you're a first time bird watcher or have been at it for a while, you will want to write everything down in this notebook to look back on and always remember your birding adventures.

9. Use it every day for writing your experiences. Also makes a great gift for Ornithologists. Keeping all your information in one spot has never been so easy. Make your memories last forever.

This Birder's journal will help you keep track of your own personal birding adventures.

Enjoy!

SEASON:

DATE: **TIME:**

LOCATION:

WEATHER/ELEMENTS

PLACE/LOCATION/HABITAT

SIGHTS/SOUNDS & ACTIVITY

BIRD SPECIES/MARKINGS/FEATURES

NOTES

SEASON:

DATE: **TIME:**

LOCATION:

WEATHER/ELEMENTS

PLACE/LOCATION/HABITAT

SIGHTS/SOUNDS & ACTIVITY

BIRD SPECIES/MARKINGS/FEATURES

NOTES

SEASON:
DATE: **TIME:**
LOCATION:

WEATHER/ELEMENTS

PLACE/LOCATION/HABITAT

SIGHTS/SOUNDS & ACTIVITY

BIRD SPECIES/MARKINGS/FEATURES

NOTES

SEASON:

DATE: **TIME:**

LOCATION:

WEATHER/ELEMENTS

PLACE/LOCATION/HABITAT

SIGHTS/SOUNDS & ACTIVITY

BIRD SPECIES/MARKINGS/FEATURES

NOTES

SEASON:

DATE: **TIME:**

LOCATION:

WEATHER/ELEMENTS

PLACE/LOCATION/HABITAT

SIGHTS/SOUNDS & ACTIVITY

BIRD SPECIES/MARKINGS/FEATURES

NOTES

SEASON:

DATE: **TIME:**

LOCATION:

WEATHER/ELEMENTS

PLACE/LOCATION/HABITAT

SIGHTS/SOUNDS & ACTIVITY

BIRD SPECIES/MARKINGS/FEATURES

NOTES

SEASON:

DATE: **TIME:**

LOCATION:

WEATHER/ELEMENTS

PLACE/LOCATION/HABITAT

SIGHTS/SOUNDS & ACTIVITY

BIRD SPECIES/MARKINGS/FEATURES

NOTES

SEASON:
DATE:　　　　　　　　　**TIME:**
LOCATION:

WEATHER/ELEMENTS

PLACE/LOCATION/HABITAT

SIGHTS/SOUNDS & ACTIVITY

BIRD SPECIES/MARKINGS/FEATURES

NOTES

SEASON:
DATE: **TIME:**
LOCATION:

WEATHER/ELEMENTS

PLACE/LOCATION/HABITAT

SIGHTS/SOUNDS & ACTIVITY

BIRD SPECIES/MARKINGS/FEATURES

NOTES

SEASON:

DATE: **TIME:**

LOCATION:

WEATHER/ELEMENTS

PLACE/LOCATION/HABITAT

SIGHTS/SOUNDS & ACTIVITY

BIRD SPECIES/MARKINGS/FEATURES

NOTES

SEASON:
DATE: **TIME:**
LOCATION:

WEATHER/ELEMENTS

PLACE/LOCATION/HABITAT

SIGHTS/SOUNDS & ACTIVITY

BIRD SPECIES/MARKINGS/FEATURES

NOTES

SEASON:

DATE: **TIME:**

LOCATION:

WEATHER/ELEMENTS

PLACE/LOCATION/HABITAT

SIGHTS/SOUNDS & ACTIVITY

BIRD SPECIES/MARKINGS/FEATURES

NOTES

SEASON:

DATE: **TIME:**

LOCATION:

WEATHER/ELEMENTS

PLACE/LOCATION/HABITAT

SIGHTS/SOUNDS & ACTIVITY

BIRD SPECIES/MARKINGS/FEATURES

NOTES

SEASON:
DATE: **TIME:**
LOCATION:

WEATHER/ELEMENTS

PLACE/LOCATION/HABITAT

SIGHTS/SOUNDS & ACTIVITY

BIRD SPECIES/MARKINGS/FEATURES

NOTES

SEASON:
DATE: **TIME:**
LOCATION:

WEATHER/ELEMENTS

PLACE/LOCATION/HABITAT

SIGHTS/SOUNDS & ACTIVITY

BIRD SPECIES/MARKINGS/FEATURES

NOTES

SEASON:

DATE: **TIME:**

LOCATION:

WEATHER/ELEMENTS

PLACE/LOCATION/HABITAT

SIGHTS/SOUNDS & ACTIVITY

BIRD SPECIES/MARKINGS/FEATURES

NOTES

SEASON:

DATE: **TIME:**

LOCATION:

WEATHER/ELEMENTS

PLACE/LOCATION/HABITAT

SIGHTS/SOUNDS & ACTIVITY

BIRD SPECIES/MARKINGS/FEATURES

NOTES

SEASON:
DATE: **TIME:**
LOCATION:

WEATHER/ELEMENTS

PLACE/LOCATION/HABITAT

SIGHTS/SOUNDS & ACTIVITY

_____

_____

_____

BIRD SPECIES/MARKINGS/FEATURES

_____

_____

NOTES

_____

_____

_____

_____

_____

SEASON:
DATE: **TIME:**
LOCATION:

WEATHER/ELEMENTS
PLACE/LOCATION/HABITAT

SIGHTS/SOUNDS & ACTIVITY

BIRD SPECIES/MARKINGS/FEATURES

NOTES

SEASON:

DATE: **TIME:**

LOCATION:

WEATHER/ELEMENTS

PLACE/LOCATION/HABITAT

SIGHTS/SOUNDS & ACTIVITY

BIRD SPECIES/MARKINGS/FEATURES

NOTES

SEASON:

DATE: **TIME:**

LOCATION:

WEATHER/ELEMENTS

PLACE/LOCATION/HABITAT

SIGHTS/SOUNDS & ACTIVITY

BIRD SPECIES/MARKINGS/FEATURES

NOTES

SEASON:

DATE: **TIME:**

LOCATION:

WEATHER/ELEMENTS
PLACE/LOCATION/HABITAT

SIGHTS/SOUNDS & ACTIVITY

BIRD SPECIES/MARKINGS/FEATURES

NOTES

SEASON:

DATE: **TIME:**

LOCATION:

WEATHER/ELEMENTS

PLACE/LOCATION/HABITAT

SIGHTS/SOUNDS & ACTIVITY

BIRD SPECIES/MARKINGS/FEATURES

NOTES

SEASON:
DATE: **TIME:**
LOCATION:

WEATHER/ELEMENTS

PLACE/LOCATION/HABITAT

SIGHTS/SOUNDS & ACTIVITY

BIRD SPECIES/MARKINGS/FEATURES

NOTES

SEASON:
DATE: **TIME:**
LOCATION:

WEATHER/ELEMENTS

PLACE/LOCATION/HABITAT

SIGHTS/SOUNDS & ACTIVITY

BIRD SPECIES/MARKINGS/FEATURES

NOTES

SEASON:
DATE: **TIME:**
LOCATION:

WEATHER/ELEMENTS

PLACE/LOCATION/HABITAT

SIGHTS/SOUNDS & ACTIVITY

BIRD SPECIES/MARKINGS/FEATURES

NOTES

SEASON:

DATE:　　　　　　　　　　　**TIME:**

LOCATION:

WEATHER/ELEMENTS

PLACE/LOCATION/HABITAT

SIGHTS/SOUNDS & ACTIVITY

BIRD SPECIES/MARKINGS/FEATURES

NOTES

SEASON:

DATE: **TIME:**

LOCATION:

WEATHER/ELEMENTS

PLACE/LOCATION/HABITAT

SIGHTS/SOUNDS & ACTIVITY

BIRD SPECIES/MARKINGS/FEATURES

NOTES

SEASON:
DATE: **TIME:**
LOCATION:

WEATHER/ELEMENTS

PLACE/LOCATION/HABITAT

SIGHTS/SOUNDS & ACTIVITY

BIRD SPECIES/MARKINGS/FEATURES

NOTES

SEASON:
DATE: **TIME:**
LOCATION:

WEATHER/ELEMENTS

PLACE/LOCATION/HABITAT

SIGHTS/SOUNDS & ACTIVITY

BIRD SPECIES/MARKINGS/FEATURES

NOTES

SEASON:

DATE: **TIME:**

LOCATION:

WEATHER/ELEMENTS

PLACE/LOCATION/HABITAT

SIGHTS/SOUNDS & ACTIVITY

BIRD SPECIES/MARKINGS/FEATURES

NOTES

SEASON:
DATE: **TIME:**
LOCATION:

WEATHER/ELEMENTS

PLACE/LOCATION/HABITAT

SIGHTS/SOUNDS & ACTIVITY

BIRD SPECIES/MARKINGS/FEATURES

NOTES

SEASON:

DATE: TIME:

LOCATION:

WEATHER/ELEMENTS

PLACE/LOCATION/HABITAT

SIGHTS/SOUNDS & ACTIVITY

BIRD SPECIES/MARKINGS/FEATURES

NOTES

SEASON:

DATE: **TIME:**

LOCATION:

WEATHER/ELEMENTS

PLACE/LOCATION/HABITAT

SIGHTS/SOUNDS & ACTIVITY

BIRD SPECIES/MARKINGS/FEATURES

NOTES

SEASON:
DATE: **TIME:**
LOCATION:

WEATHER/ELEMENTS
PLACE/LOCATION/HABITAT

SIGHTS/SOUNDS & ACTIVITY

BIRD SPECIES/MARKINGS/FEATURES

NOTES

SEASON:
DATE: **TIME:**
LOCATION:

WEATHER/ELEMENTS

PLACE/LOCATION/HABITAT

SIGHTS/SOUNDS & ACTIVITY

BIRD SPECIES/MARKINGS/FEATURES

NOTES

SEASON:

DATE: **TIME:**

LOCATION:

WEATHER/ELEMENTS

PLACE/LOCATION/HABITAT

SIGHTS/SOUNDS & ACTIVITY

BIRD SPECIES/MARKINGS/FEATURES

NOTES

SEASON:
DATE:　　　　　　　　**TIME:**
LOCATION:

WEATHER/ELEMENTS

PLACE/LOCATION/HABITAT

SIGHTS/SOUNDS & ACTIVITY

BIRD SPECIES/MARKINGS/FEATURES

NOTES

SEASON:
DATE: **TIME:**
LOCATION:

WEATHER/ELEMENTS

PLACE/LOCATION/HABITAT

SIGHTS/SOUNDS & ACTIVITY

BIRD SPECIES/MARKINGS/FEATURES

NOTES

SEASON:

DATE: **TIME:**

LOCATION:

WEATHER/ELEMENTS

PLACE/LOCATION/HABITAT

SIGHTS/SOUNDS & ACTIVITY

BIRD SPECIES/MARKINGS/FEATURES

NOTES

SEASON:
DATE: **TIME:**
LOCATION:

WEATHER/ELEMENTS

PLACE/LOCATION/HABITAT

SIGHTS/SOUNDS & ACTIVITY

BIRD SPECIES/MARKINGS/FEATURES

NOTES

SEASON:

DATE: **TIME:**

LOCATION:

WEATHER/ELEMENTS

PLACE/LOCATION/HABITAT

SIGHTS/SOUNDS & ACTIVITY

BIRD SPECIES/MARKINGS/FEATURES

NOTES

SEASON:

DATE: **TIME:**

LOCATION:

WEATHER/ELEMENTS

PLACE/LOCATION/HABITAT

SIGHTS/SOUNDS & ACTIVITY

BIRD SPECIES/MARKINGS/FEATURES

NOTES

SEASON:
DATE: **TIME:**
LOCATION:

WEATHER/ELEMENTS

PLACE/LOCATION/HABITAT

SIGHTS/SOUNDS & ACTIVITY

BIRD SPECIES/MARKINGS/FEATURES

NOTES

SEASON:
DATE: **TIME:**
LOCATION:

WEATHER/ELEMENTS

PLACE/LOCATION/HABITAT

SIGHTS/SOUNDS & ACTIVITY

BIRD SPECIES/MARKINGS/FEATURES

NOTES

SEASON:
DATE: **TIME:**
LOCATION:

WEATHER/ELEMENTS

PLACE/LOCATION/HABITAT

SIGHTS/SOUNDS & ACTIVITY

BIRD SPECIES/MARKINGS/FEATURES

NOTES

SEASON:

DATE: **TIME:**

LOCATION:

WEATHER/ELEMENTS

PLACE/LOCATION/HABITAT

SIGHTS/SOUNDS & ACTIVITY

BIRD SPECIES/MARKINGS/FEATURES

NOTES

SEASON:
DATE: **TIME:**
LOCATION:

WEATHER/ELEMENTS

PLACE/LOCATION/HABITAT

SIGHTS/SOUNDS & ACTIVITY

BIRD SPECIES/MARKINGS/FEATURES

NOTES

SEASON:
DATE: **TIME:**
LOCATION:

WEATHER/ELEMENTS

PLACE/LOCATION/HABITAT

SIGHTS/SOUNDS & ACTIVITY

BIRD SPECIES/MARKINGS/FEATURES

NOTES

SEASON:
DATE: **TIME:**
LOCATION:

WEATHER/ELEMENTS

PLACE/LOCATION/HABITAT

SIGHTS/SOUNDS & ACTIVITY

BIRD SPECIES/MARKINGS/FEATURES

NOTES

SEASON:
DATE: **TIME:**
LOCATION:

WEATHER/ELEMENTS

PLACE/LOCATION/HABITAT

SIGHTS/SOUNDS & ACTIVITY

BIRD SPECIES/MARKINGS/FEATURES

NOTES

SEASON:
DATE: **TIME:**
LOCATION:

WEATHER/ELEMENTS

PLACE/LOCATION/HABITAT

SIGHTS/SOUNDS & ACTIVITY

BIRD SPECIES/MARKINGS/FEATURES

NOTES

SEASON:

DATE: **TIME:**

LOCATION:

WEATHER/ELEMENTS

PLACE/LOCATION/HABITAT

SIGHTS/SOUNDS & ACTIVITY

BIRD SPECIES/MARKINGS/FEATURES

NOTES

SEASON:
DATE: **TIME:**
LOCATION:

WEATHER/ELEMENTS

PLACE/LOCATION/HABITAT

SIGHTS/SOUNDS & ACTIVITY

BIRD SPECIES/MARKINGS/FEATURES

NOTES

SEASON:

DATE: **TIME:**

LOCATION:

WEATHER/ELEMENTS

PLACE/LOCATION/HABITAT

SIGHTS/SOUNDS & ACTIVITY

BIRD SPECIES/MARKINGS/FEATURES

NOTES

SEASON:
DATE: **TIME:**
LOCATION:

WEATHER/ELEMENTS

PLACE/LOCATION/HABITAT

SIGHTS/SOUNDS & ACTIVITY

BIRD SPECIES/MARKINGS/FEATURES

NOTES

SEASON:

DATE: **TIME:**

LOCATION:

WEATHER/ELEMENTS

PLACE/LOCATION/HABITAT

SIGHTS/SOUNDS & ACTIVITY

BIRD SPECIES/MARKINGS/FEATURES

NOTES

SEASON:
DATE: **TIME:**
LOCATION:

WEATHER/ELEMENTS

PLACE/LOCATION/HABITAT

SIGHTS/SOUNDS & ACTIVITY

BIRD SPECIES/MARKINGS/FEATURES

NOTES

SEASON:

DATE: **TIME:**

LOCATION:

WEATHER/ELEMENTS

PLACE/LOCATION/HABITAT

SIGHTS/SOUNDS & ACTIVITY

BIRD SPECIES/MARKINGS/FEATURES

NOTES

SEASON:
DATE: **TIME:**
LOCATION:

WEATHER/ELEMENTS

PLACE/LOCATION/HABITAT

SIGHTS/SOUNDS & ACTIVITY

BIRD SPECIES/MARKINGS/FEATURES

NOTES

SEASON:
DATE: **TIME:**
LOCATION:

WEATHER/ELEMENTS

PLACE/LOCATION/HABITAT

SIGHTS/SOUNDS & ACTIVITY

BIRD SPECIES/MARKINGS/FEATURES

NOTES

SEASON:

DATE: **TIME:**

LOCATION:

WEATHER/ELEMENTS

PLACE/LOCATION/HABITAT

SIGHTS/SOUNDS & ACTIVITY

BIRD SPECIES/MARKINGS/FEATURES

NOTES

SEASON:

DATE: **TIME:**

LOCATION:

WEATHER/ELEMENTS

PLACE/LOCATION/HABITAT

SIGHTS/SOUNDS & ACTIVITY

BIRD SPECIES/MARKINGS/FEATURES

NOTES

SEASON:
DATE: **TIME:**
LOCATION:

WEATHER/ELEMENTS

PLACE/LOCATION/HABITAT

SIGHTS/SOUNDS & ACTIVITY

BIRD SPECIES/MARKINGS/FEATURES

NOTES

SEASON:
DATE: **TIME:**
LOCATION:

WEATHER/ELEMENTS

PLACE/LOCATION/HABITAT

SIGHTS/SOUNDS & ACTIVITY

BIRD SPECIES/MARKINGS/FEATURES

NOTES

SEASON:
DATE: **TIME:**
LOCATION:

WEATHER/ELEMENTS

PLACE/LOCATION/HABITAT

SIGHTS/SOUNDS & ACTIVITY

BIRD SPECIES/MARKINGS/FEATURES

NOTES

SEASON:
DATE: **TIME:**
LOCATION:

WEATHER/ELEMENTS

_____

_____

PLACE/LOCATION/HABITAT

_____

_____

_____

_____

SIGHTS/SOUNDS & ACTIVITY

_____

_____

_____

BIRD SPECIES/MARKINGS/FEATURES

_____

_____

_____

NOTES

_____

SEASON:
DATE: **TIME:**
LOCATION:

WEATHER/ELEMENTS

PLACE/LOCATION/HABITAT

SIGHTS/SOUNDS & ACTIVITY

BIRD SPECIES/MARKINGS/FEATURES

NOTES

SEASON:
DATE: **TIME:**
LOCATION:

WEATHER/ELEMENTS

PLACE/LOCATION/HABITAT

SIGHTS/SOUNDS & ACTIVITY

BIRD SPECIES/MARKINGS/FEATURES

NOTES

SEASON:
DATE: **TIME:**
LOCATION:

WEATHER/ELEMENTS

PLACE/LOCATION/HABITAT

SIGHTS/SOUNDS & ACTIVITY

BIRD SPECIES/MARKINGS/FEATURES

NOTES

SEASON:
DATE: **TIME:**
LOCATION:

WEATHER/ELEMENTS

PLACE/LOCATION/HABITAT

SIGHTS/SOUNDS & ACTIVITY

BIRD SPECIES/MARKINGS/FEATURES

NOTES

SEASON:

DATE: **TIME:**

LOCATION:

WEATHER/ELEMENTS

PLACE/LOCATION/HABITAT

SIGHTS/SOUNDS & ACTIVITY

BIRD SPECIES/MARKINGS/FEATURES

NOTES

SEASON:
DATE: **TIME:**
LOCATION:

WEATHER/ELEMENTS

PLACE/LOCATION/HABITAT

SIGHTS/SOUNDS & ACTIVITY

BIRD SPECIES/MARKINGS/FEATURES

NOTES

SEASON:

DATE: **TIME:**

LOCATION:

WEATHER/ELEMENTS

PLACE/LOCATION/HABITAT

SIGHTS/SOUNDS & ACTIVITY

BIRD SPECIES/MARKINGS/FEATURES

NOTES

SEASON:
DATE: **TIME:**
LOCATION:

WEATHER/ELEMENTS

PLACE/LOCATION/HABITAT

SIGHTS/SOUNDS & ACTIVITY

BIRD SPECIES/MARKINGS/FEATURES

NOTES

SEASON:

DATE: **TIME:**

LOCATION:

WEATHER/ELEMENTS

PLACE/LOCATION/HABITAT

SIGHTS/SOUNDS & ACTIVITY

BIRD SPECIES/MARKINGS/FEATURES

NOTES

SEASON:
DATE: **TIME:**
LOCATION:

WEATHER/ELEMENTS

PLACE/LOCATION/HABITAT

SIGHTS/SOUNDS & ACTIVITY

BIRD SPECIES/MARKINGS/FEATURES

NOTES

SEASON:
DATE: **TIME:**
LOCATION:

WEATHER/ELEMENTS

PLACE/LOCATION/HABITAT

SIGHTS/SOUNDS & ACTIVITY

BIRD SPECIES/MARKINGS/FEATURES

NOTES

WEATHER/ELEMENTS

PLACE/LOCATION/HABITAT

SIGHTS/SOUNDS & ACTIVITY

BIRD SPECIES/MARKINGS/FEATURES

NOTES

SEASON:

DATE: **TIME:**

LOCATION:

WEATHER/ELEMENTS

PLACE/LOCATION/HABITAT

SIGHTS/SOUNDS & ACTIVITY

BIRD SPECIES/MARKINGS/FEATURES

NOTES

SEASON:
DATE: **TIME:**
LOCATION:

WEATHER/ELEMENTS

PLACE/LOCATION/HABITAT

SIGHTS/SOUNDS & ACTIVITY

BIRD SPECIES/MARKINGS/FEATURES

NOTES

SEASON:
DATE: **TIME:**
LOCATION:

WEATHER/ELEMENTS

PLACE/LOCATION/HABITAT

SIGHTS/SOUNDS & ACTIVITY

BIRD SPECIES/MARKINGS/FEATURES

NOTES

SEASON:
DATE: **TIME:**
LOCATION:

WEATHER/ELEMENTS

PLACE/LOCATION/HABITAT

SIGHTS/SOUNDS & ACTIVITY

BIRD SPECIES/MARKINGS/FEATURES

NOTES

SEASON:
DATE: **TIME:**
LOCATION:

WEATHER/ELEMENTS

PLACE/LOCATION/HABITAT

SIGHTS/SOUNDS & ACTIVITY

BIRD SPECIES/MARKINGS/FEATURES

NOTES

SEASON:
DATE: **TIME:**
LOCATION:

WEATHER/ELEMENTS

PLACE/LOCATION/HABITAT

SIGHTS/SOUNDS & ACTIVITY

BIRD SPECIES/MARKINGS/FEATURES

NOTES

SEASON:
DATE: **TIME:**
LOCATION:

WEATHER/ELEMENTS

PLACE/LOCATION/HABITAT

SIGHTS/SOUNDS & ACTIVITY

BIRD SPECIES/MARKINGS/FEATURES

NOTES

SEASON:

DATE: **TIME:**

LOCATION:

WEATHER/ELEMENTS

PLACE/LOCATION/HABITAT

SIGHTS/SOUNDS & ACTIVITY

BIRD SPECIES/MARKINGS/FEATURES

NOTES

SEASON:
DATE: **TIME:**
LOCATION:

WEATHER/ELEMENTS

PLACE/LOCATION/HABITAT

SIGHTS/SOUNDS & ACTIVITY

BIRD SPECIES/MARKINGS/FEATURES

NOTES

SEASON:
DATE: **TIME:**
LOCATION:

WEATHER/ELEMENTS

PLACE/LOCATION/HABITAT

SIGHTS/SOUNDS & ACTIVITY

BIRD SPECIES/MARKINGS/FEATURES

NOTES

SEASON:
DATE: **TIME:**
LOCATION:

WEATHER/ELEMENTS

PLACE/LOCATION/HABITAT

SIGHTS/SOUNDS & ACTIVITY

BIRD SPECIES/MARKINGS/FEATURES

NOTES

SEASON:

DATE: TIME:

LOCATION:

WEATHER/ELEMENTS

PLACE/LOCATION/HABITAT

SIGHTS/SOUNDS & ACTIVITY

BIRD SPECIES/MARKINGS/FEATURES

NOTES

SEASON:
DATE: **TIME:**
LOCATION:

WEATHER/ELEMENTS

PLACE/LOCATION/HABITAT

SIGHTS/SOUNDS & ACTIVITY

BIRD SPECIES/MARKINGS/FEATURES

NOTES

SEASON:
DATE: **TIME:**
LOCATION:

WEATHER/ELEMENTS

PLACE/LOCATION/HABITAT

SIGHTS/SOUNDS & ACTIVITY

BIRD SPECIES/MARKINGS/FEATURES

NOTES

SEASON:

DATE: **TIME:**

LOCATION:

WEATHER/ELEMENTS

PLACE/LOCATION/HABITAT

SIGHTS/SOUNDS & ACTIVITY

BIRD SPECIES/MARKINGS/FEATURES

NOTES

SEASON:

DATE: **TIME:**

LOCATION:

WEATHER/ELEMENTS

PLACE/LOCATION/HABITAT

SIGHTS/SOUNDS & ACTIVITY

BIRD SPECIES/MARKINGS/FEATURES

NOTES

SEASON:

DATE: **TIME:**

LOCATION:

WEATHER/ELEMENTS

PLACE/LOCATION/HABITAT

SIGHTS/SOUNDS & ACTIVITY

BIRD SPECIES/MARKINGS/FEATURES

NOTES

SEASON:
DATE: **TIME:**
LOCATION:

WEATHER/ELEMENTS

PLACE/LOCATION/HABITAT

SIGHTS/SOUNDS & ACTIVITY

BIRD SPECIES/MARKINGS/FEATURES

NOTES

SEASON:
DATE: **TIME:**
LOCATION:

WEATHER/ELEMENTS

PLACE/LOCATION/HABITAT

SIGHTS/SOUNDS & ACTIVITY

BIRD SPECIES/MARKINGS/FEATURES

NOTES

SEASON:

DATE: **TIME:**

LOCATION:

WEATHER/ELEMENTS

PLACE/LOCATION/HABITAT

SIGHTS/SOUNDS & ACTIVITY

BIRD SPECIES/MARKINGS/FEATURES

NOTES

SEASON:

DATE: **TIME:**

LOCATION:

WEATHER/ELEMENTS

PLACE/LOCATION/HABITAT

SIGHTS/SOUNDS & ACTIVITY

BIRD SPECIES/MARKINGS/FEATURES

NOTES

SEASON:
DATE: **TIME:**
LOCATION:

WEATHER/ELEMENTS

PLACE/LOCATION/HABITAT

SIGHTS/SOUNDS & ACTIVITY

BIRD SPECIES/MARKINGS/FEATURES

NOTES

SEASON:
DATE: **TIME:**
LOCATION:

WEATHER/ELEMENTS

PLACE/LOCATION/HABITAT

SIGHTS/SOUNDS & ACTIVITY

BIRD SPECIES/MARKINGS/FEATURES

NOTES

SEASON:
DATE: **TIME:**
LOCATION:

WEATHER/ELEMENTS
PLACE/LOCATION/HABITAT

SIGHTS/SOUNDS & ACTIVITY

BIRD SPECIES/MARKINGS/FEATURES

NOTES

SEASON:
DATE: **TIME:**
LOCATION:

WEATHER/ELEMENTS

PLACE/LOCATION/HABITAT

SIGHTS/SOUNDS & ACTIVITY

BIRD SPECIES/MARKINGS/FEATURES

NOTES

SEASON:
DATE: **TIME:**
LOCATION:

WEATHER/ELEMENTS

PLACE/LOCATION/HABITAT

SIGHTS/SOUNDS & ACTIVITY

BIRD SPECIES/MARKINGS/FEATURES

NOTES

SEASON:
DATE: **TIME:**
LOCATION:

WEATHER/ELEMENTS
PLACE/LOCATION/HABITAT

SIGHTS/SOUNDS & ACTIVITY

BIRD SPECIES/MARKINGS/FEATURES

NOTES

WEATHER/ELEMENTS

PLACE/LOCATION/HABITAT

SIGHTS/SOUNDS & ACTIVITY

BIRD SPECIES/MARKINGS/FEATURES

NOTES

Lightning Source UK Ltd.
Milton Keynes UK
UKHW051149310520
364201UK00007B/302